Distributors:

for the United States and Canada
Kluwer Boston, Inc.
190, Old Derby Street
Hingham, MA 02043
USA

for all other countries
Kluwer Academic Publishers Group
Distribution Center
P.O. Box 322
3300 AH Dordrecht
The Netherlands

This volume is listed in the Library of Congress Cataloging in Publication Data

ISBN-13: 978-90-247-2511-3   e-ISBN-13: 978-94-009-8317-5
DOI: 10.1007/978-94-009-8317-5

Publication arranged by:
Commission of the European Communities,
Directorate-General Information Market and Innovation,
Luxembourg

EUR 7320

Copyright © ECSC, EEC, EAEC, Brussels-Luxembourg, 1981

All rights reserved. No part of this publication may be reproduced, stored in a retrieval system, or transmitted in any form or by any means, mechanical, photocopying, recording or otherwise, without the prior written permission of the copyright owners.
Martinus Nijhoff Publishers bv, P.O. Box 566, 2501 CN The Hague, The Netherlands.

LEGAL NOTICE

Neither the commission of the European Communities nor any person acting on behalf of the Commission is responsible for the use which might be made of the following information.

Achievements of The European Community First Energy R & D Programme

by

J. T. McMullan,
New University of Ulster, Coleraine, Northern Ireland

and

A. S. Strub,
Commission of the European Community, Brussels

1981

MARTINUS NIJHOFF PUBLISHERS
THE HAGUE / BOSTON / LONDON
for
THE COMMISSION OF THE EUROPEAN COMMUNITIES

Eurelios solar furnace, Sicily. (Consortium consisting of ENEL/ANSALDO, Italy, CETHEL, France and MBB, W. Germany).

CONTENTS

THE EUROPEAN COMMUNITY ENERGY R & D PROGRAMMES	5
ENERGY CONSERVATION	7
Energy Savings in Buildings	7
Use of Heat Pumps	8
Urban Transport	10
Recovery of Residual Heat	10
Recycling of Materials	12
Production of Energy from Waste	14
Industrial Processes	14
Development of Methods of Accumulating Secondary Energy	15
Second Programme	15
PRODUCTION AND USE OF HYDROGEN	17
Thermochemical Production of Hydrogen	17
Electrolytic Production of Hydrogen	20
Hydrogen Use, Storage and Transportation	20
Second Programme	22
SOLAR ENERGY	23
Solar Energy Applications to Dwellings	24
Thermomechanical Solar Power Plants	25
Photovoltaic Power Generation	25
Photoelectrochemical, Photochemical and Photobiological Processes	28
Energy from Biomass	29
Solar Radiation Data	29
Second Programme	32
GEOTHERMAL ENERGY	33
Acquisition and Collection of Existing and New Geothermal Data	33
Prospecting Methodology	34
Utilisation of Hot Water Sources (Low Enthalpy)	36
Utilisation of Steam Sources (High Enthalpy)	37
Hot Dry Rocks	37
Second Programme	38
ENERGY SYSTEMS ANALYSIS AND STRATEGY STUDIES	39
First Programme	41
Second Programme	44
CONCLUSIONS	45
APPENDIX 1 — Conferences and Publications arising out of the First Energy R & D Programme	47

Blasting (explosive stimulation) at hot dry rocks test site in Cornwall (Cambourne School of Mines, United Kingdom).

THE EUROPEAN COMMUNITY ENERGY R & D PROGRAMMES

The European Community has been involved in certain fields of energy research for a considerable length of time. Within the framework of the European Coal and Steel Treaty (1953), coal research has been carried out for over twenty five years. In the nuclear energy area, the European Atomic Energy Community was created in 1958, and since then, research has been carried out in many fields related to both nuclear fission energy and thermonuclear fusion. Since the oil crisis of 1973-74, these research activities have been reinforced and complemented by efforts in other areas. In particular, the impetus provided by the 1973-74 crisis and the subsequent price rises stimulated awareness of the importance of renewable energy supplies and energy conservation, and one of the Community responses was to launch what was called the First Energy R & D Programme in 1975, with a duration of four years. This was followed by the Second Energy R & D Programme in 1979. Both of these programmes are of the Indirect Action type, and are aimed at stimulating contract research in the renewable energy and energy conservation areas, with the Commission paying up to 50% of the total research cost. Other responses were also made under the programme of the Joint Research Centre, through the financial support of large scale Demonstration Projects, and through an extension of the thermonuclear fusion programme in the separately funded JET project.

To show the measure of the EEC commitment to Energy R & D, it is worth noting that, of a total European expenditure of over 2,500 MioECU per year*, the Commission contributes about 10%. (This should be compared with 2% for "general" R & D). Thus, it is one of the major moving forces in the European energy research effort. Indeed, in certain areas, the contribution of the Commission is greater — 30% for fusion research, and 40% for the production and use of hydrogen. Also, energy topics account for 70% of the total Community research budget.

The first Energy R & D Programme is now complete and has proved to be highly successful. With the initial stimulus of CREST-ENERGY (the Energy sub-committee of CREST, the "Comité de la Recherche Scientifique et Technique") and through its efforts and those of the Advisory Committees on Project Management (ACPM's), an extremely large number of independent, but collaborating, contractors and projects have been supported, with results that are to the benefit of the European Community as a whole. The purpose of this document is to present an overall picture of the aims and achievements of this programme. The objectives of the second programme will also be presented since these were formulated in the light of the experience gained in the first phase.

The first Energy R & D Programme is divided into five Sub-programmes concerned with Energy Conservation, The Production and Use of Hydrogen, Solar Energy, Geothermal Energy, and Energy Systems Analysis and Strategy. Each of these will be dealt with separately, but some features are common to all of them. These include the system of programme management and control, and the steps taken to ensure that — to as great a degree as possible, bearing in mind the need to protect proprietary interests — information on developments made under the ægis of the programme is disseminated as widely as possible throughout the Community. Thus, all projects must provide for regular progress reports and a final report which is suitable for publication. Progress is monitored by an expert in the particular field who is responsible to the Commission. Regular contractors meetings are held to ensure that there is adequate interaction between project leaders in related areas, and collaborative projects are actively encouraged. In addition, international conferences are held to encourage the dissemination of information to a wider audience, and to induce collaboration or interaction between wider groups of workers. A complete list of the conferences organised under the Energy R & D Programme is given in Appendix 1.

It is worth noting that the open tender contract research system of R & D support used for the programme, which is the method most widely used by the Commission for supporting Energy R & D, has many attractions and advantages. For example, all of the skills and expertise available throughout the member states is accessible, whereas much might well be undetected by more direct sponsorship methods. Another important factor is the way in which work can be co-ordinated and collaboration established between laboratories with related interests. Frequently, similar

---

*MioECU = Million European Currency Units
At February 28th 1981, 1ECU = 41.67 BF = 7.99 DKR = 2.59 DM = 60.93 DRA = 5.99 FF = 0.70 IR£ = 1231 LIT = 2.81 HFL = 0.52 UK£ = 1.23 US$.

proposals appear from two or more institutions, which, as likely as not, will be in different member states. In such cases, collaboration between the groups can be actively stimulated by the Commission. Effectively, the work is supported once instead of twice, but both centres are involved and facilities are included in the final contract to ensure that useful collaboration occurs. The benefits of this collaboration have been evident throughout the programme.

The Council Decisions to approve the two four year Energy R & D Programmes were taken on 22 Aug. 1975 for the First Programme and 11 September 1979 for the Second. They were subsequently approved by the European Parliament. The total budget for the First Programme was 59 MioECU, and for the Second Programme 105 MioECU. The distributions between the five sub-programmes are as shown in Tables 1 and 2, where it can be seen that, with the exception of Production and Use of Hydrogen, the expenditure in the Second Programme was increased in all areas. In the case of the hydrogen sub-programme, the reduced funding is the result of recognising that hydrogen will make primarily a long term contribution; as a result, there has been a change of emphasis towards improving electrolytic production methods and developing applications studies.

For a number of reasons, the overall increase in activity is not obvious in the number of projects supported. These include the introduction of pilot projects, the increased scale and consequent increased cost of a number of the other projects, and the fact that the selection process is not complete at the time of writing. This means that the second column of Table 2 should be treated with caution and not read as a representation of the total second programme.

The remainder of this document is devoted to a discussion of the individual sub-programmes. The same format will be used for each one; a brief introduction, a statement of the aims of the sub-programme and the study areas, or projects, into which it is divided, then a statement of the objectives and achievements of each of the projects. Finally, a short summary of the aims of the corresponding part of the Second Programme is given. It is hoped that in this way, an overall picture of the successes — and failures — of the First Programme can be appreciated.

	Budget (MioECU)	No. of Projects
Energy Conservation	11.38	117
Production and use of Hydrogen	13.24	83
Solar Energy	17.50	289
Geothermal Energy	13.00	140
Systems Analysis/Energy Modelling	3.88	54
TOTALS	59.00	683

Table 1. Expenditure and number of projects supported under the First Energy R & D Programme.

	Budget (MioECU)	No. of Projects (Provisional)
Energy Conservation	27	160
Production and Use of Hydrogen	8	35
Solar Energy	46	186
Geothermal Energy	18	63
Systems Analysis/Energy Modelling	6	32
TOTALS	105	444*

*Provisional total, subject to amendment as the selection procedure develops.

Table 2. Expenditure approved and number of contracts supported under the Second Energy R & D Programme.

ENERGY CONSERVATION

The first plank of Community Energy policy is to reduce specific energy consumption. That is, to reduce the energy used per unit of production, or per head of population, or whatever other basis is chosen. There is only one way of achieving this aim — *energy conservation*. All other elements of the Energy R & D Programme are aimed at finding ways of meeting the demand, only this one is aimed at actually reducing it.

From an R & D viewpoint, Energy Conservation is a very difficult field. Firstly, it is extremely wide ranging in the type of activity that can be included. Secondly, it frequently does not actually involve R & D at all, but rather the co-ordination and application of techniques which are already known. Finally, it depends for its success on the combined efforts of a large number of individuals who may or may not choose to collaborate in achieving a measure of energy conservation.

For the European Community as a whole, in 1978, the energy consumptions in the domestic, industrial and transport sectors represented approximately 40%, 40% and 20% of the total final consumption respectively, in terms of oil, the corresponding figures were 40%, 24% and 36% respectively. It is clear therefore, that no one sector can be singled out as *the one* upon which to concentrate, and that energy savings must be made in all sectors. Unfortunately, the cheap oil era of the sixties led to many industrial and domestic practices that make readjustment difficult today.

Sumary and Achievements of the First Programme (11.38 MioECU)

Eight study areas (projects) were identified in the first programme:
1. Energy savings in Buildings
2. Use of Heat Pumps
3. Urban Transport
4. Recovery of Residual Heat
5. Recycling of materials
6. Production of Energy from Waste
7. Industrial Processes
8. Development of methods for accumulating secondary energy.

Project 1. Energy Savings in Buildings

Objectives:
(a) To examine new processes for the fabrication of transparent materials which have low thermal conductivity, or are opaque in the infrared region of the spectrum.
(b) To examine new insulating materials and fabrication processes.
(c) To undertake optimisation studies to examine the combined problems of solar energy, air infiltration and heating systems.

Achievements:
1. Progress was made in the area of selective coating of glass with Tin Oxide, Indium Oxide, gold or silver with encouraging results, although silver was found to oxidise quickly. In particular, the R & D effort has demonstrated the feasibility of coating plastic for fitting to existing windows, at economic costs (1.5 ECU/m²). This coating would make a single pane window behave like a double pane one.

2. A light, self extinguishing, insulation material with a cost of 40 ECU/m³ and a light concrete with good insulating properties have been developed.

3. A reference catalogue with thermograms of typical insulation defects in building structures has been made.

4. The achievements of the optimisation studies to date are primarily negative. It has been shown that the potential for energy saving from waste water, exhaust air and flue gases is small at 2% and 4% of total energy consumption in existing and new buildings respectively. Additionally, however, the behaviour of buildings for low cost dwellings has been quantified in a much better way than before, and it is intended that the work in this section will be brought together into a design guide which will enable architects to include an element of energy planning into their design practice. In particular, the guide will allow the energy consequences of certain design decisions to be assessed.

Project 2. Use of Heat Pumps

Objectives: Superficially, the technology of heat pumps appears to exist already, through the efforts of the refrigeration and air conditioning industries, but unfortunately, the range of operating conditions to which a heat pump might be subjected, and the continuously changing conditions to which it must respond, make it a more difficult device to understand in detail and also make the control problem less tractable. A further factor is that, presently, the capital cost is relatively high compared to other heating systems, so that there is consumer resistance to its widespread introduction unless it can be shown to be truly cost-effective.

In an approach to these problems, the EC project was given five objectives:
(a) Factfinding and precommercial testing of prototype units.
(b) Development of improved and advanced heat pumps for space heating only applications (no airconditioning duty).
(c) Development of industrial heat pumps.
(d) Studies on heat sources.
(e) Control of heat pumps.

Achievements:
1. A design study has shown that very careful attention must be paid to good design, workmanship and quality control if heat pumps are to achieve the reliability and high performance that will be required for useful market penetration in the domestic and commercial sectors. It is believed, however, that the necessary high performance and low cost are achievable.

2. An electrically driven and a gas engine driven heat pump were installed in two different apartment buildings and their performance was assessed. Both used outside air as the heat source and in both cases the performance was acceptable. The electrically driven unit was completely reliable, but showed that system optimisation is required to reach the necessary performance levels. The gas engine system produced performance figures which agreed with the predictions, but the tests showed that attention to improving reliability was required in some components.

3. A number of prototype directly fired gas or fuel oil absorption cycle heat pumps have been built and tested. Design problems were identified and corrected and some components — particularly the circulation pump — were singled out for particular attention.

4. A number of very advanced heat pump designs have been built and tested. One involves the use of a very small, turbine compressor which operates on a Rankine-Rankine cycle — with the Rankine cycle turbine driving the Rankine cycle heat pump compressor. This has shown promise, but has highlighted the problems that can occur with advanced heat pump designs.

5. A high temperature industrial heat pump producing steam at about 120°C has been developed. It was shown that with piston compressors 120°C seems to be the upper working limit because of lubrication, oil and working fluid stability problems. These have been identified as an area where more development work is needed.

6. A unit based on a mixture of refrigerants has been built and tested and has shown that the resulting range of condensing and evaporating temperatures can lead to significant advantages under certain industrial conditions.

7. An extensive study has been made of the use of soil as a heat source and agreement has been achieved between theoretical and experimental values.

8. In an application study, sewage water has been successfuly considered as the heat source for large scale heat pumps; this system fits in well with the requirements of large and medium towns.

9. In a number of contracts, the problems of frost formation on the evaporator were examined, and interesting results and methods of handling the problem were obtained.

10. Important advances have been made in the control of heat-pumps, to look after their orderly start-up and shut-down, to initiate defrost mechanisms only when required, to protect against extremes of operating conditions, to vary the capacity of the unit to meet part load conditions etc. These have been useful studies which have made a major contribution to improving the controlability of heat pumps for domestic applications.

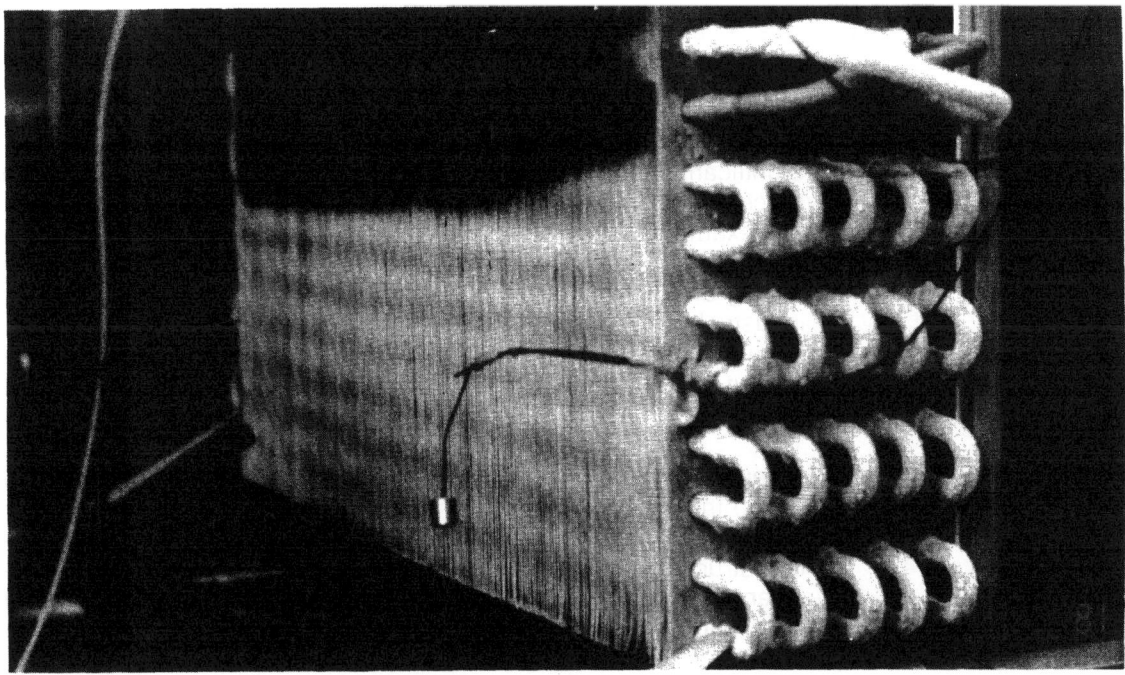

Frost formation on heat pump coils, (TNO, Netherlands).

Project 3. Urban Transport

Objectives: (a) Improvement in the efficiency of motor vehicle engines.

(b) Development of electric and hybrid cars.

Achievements:

1. Investigations of the influence of fuel-air ratio, compression ratio, and improved lubricants have shown that improvements of up to 20% in fuel economy can be achieved for petrol engines under the part load operating conditions met in urban traffic.

2. The performance of an existing petrol engine was improved by increasing the compression ratio, using lean fuel/air mixtures, controlling the ignition timing and changing the combustion chamber geometry, to such an extent that its fuel consumption was as low as that of a similar Diesel engine, though the power output was twice as high.

3. On the negative side, attempts to increase the efficiency of Diesel engines by introducing a special prechamber and by the use of ceramic pistons and liners, did not fulfil the expectations held for them.

4. A number of scenario studies, taking environmental constraints into account, have indicated that fuel savings of almost 20% are achievable by switching from Diesel and petrol engines to lean burn and stratified charge units. These are also capable of using the likely gasoline subsitutes, methanol and synthetic gasoline. The model used here could be of help to authorities attempting to define a policy in this area.

Project 4. Recovery of Residual Heat

Objectives:

(a) The development of equipment to permit the recovery of heat which is normally wasted (hot water, flue gases, etc.)

(b) The development of methods of transporting recovered heat.

(c) The development of combined recovery/utilisation systems.

(d) The development of filters for high temperature gases.

Achievements:

1. Significant energy savings have been achieved in different industrial processes by redesign so that waste heat is recovered and reused. For example, 60% usable heat recovery from the flue gases in a scrap-iron melting plant led to a 1 year payback period; a 25% energy saving and a 40% productivity increase was achieved in a glass industry by recovery of flue gas heat to preheat the powder charge. Other examples were found in the food industry and the coke industry but the aluminium industry was found to offer only a limited potential.

2. Extensive studies were made of gravity assisted heat pipes and two phase thermosyphons both of which were found to have excellent heat transfer properties, and the long term compatibility of the various structural options was found to be good. The agreement between practical measurements of the heat transfer and theoretical calculations was not good, however, and leaves scope for improvement.

Rankine Cycle Engine: Expander (left; turbine wheel, right; feeding volute and nozzle). (Fiat, Italy).

Rankine Cycle Engine: Turbine Wheel, (right: sample with shroud removed).

3. An extensive study of the use of a thermal wheel (rotating regenerator) in a synthetic fibre drying oven highlighted the difficulties to be encountered in trying to retrofit such equipment to existing installations. For example, because of the open ended design of the existing dryer, it was difficult to prevent excessive stray leakage and to direct the hot exhaust to the heat recovery devices. The duct work for such applications proved to be too expensive to allow retrofitting to be economic.

4. When waste heat is rejected at 250°C or above it is tempting to consider using it to produce electricity directly by a Rankine cycle turbine. Two such studies have been made. In the first, refrigerant R11 was used as the working fluid in an engine using 250°C waste heat at the inlet. Efficiencies of 15% were achieved, but serious corrosion problems were encountered because of the decomposition of R11 at temperatures above 150°C. Changing to the more stable perfluorohexane, the engine performed well up to 250°C but because the engine was not designed for this fluid the power output was reduced from 95 to 37kW and the efficiency to 7%. Perfluorohexane would require a much more expensive engine and is itself an expensive material.

In the second, exhaust gases from ceramic tunnel ovens were available at temperatures of 250°C. This is potentially a large source of power if suitable uses can be found. The Rankine cycle generator was based on tetrachloroethylene as the working fluid and performed well, however the calculated power (40kW) and efficiency (11%) were not achieved because the exhaust gases had cooled from 250°C to 175°C in the duct before reaching the engine. This fault is now being remedied. A payback period of 3 years is estimated.

Project 5 Recycling of Materials

Objectives:

(a) To make a detailed study to determine the usefulness of the different recycling possibilities, their economic return and the possibilities for using the recycled products.

(b) To develop appropriate recycling processes.

Achievements:

1. An industrial plant has been designed for recycling domestic mixed plastic waste on a scale of 2,400 T/yr, and construction is to proceed with the sponsorship of the EEC. (Demonstration project). The technique is a new one in which the mixed plastic waste is ground, washed and dried and then transformed into a homogenous plastic product which is then crushed into granules. These granules are used as feed materials for injection moulding, extrusion and blow moulding extrusion of plastic products.

2. A method has been developed for recovering the glass fibres from reinforced thermoset resins in which heat is used to destroy the polymer after grinding, leaving the fibres which can then be used to reinforce thermoplastic wastes to improve their mechanical properties. It seems likely that in a plant of 250 T/yr capacity, with a glass output of 100 T/yr, the payback period would be about 3 years.

Experimental Heat Pump operating with a mixture of refrigerants, (IFP, France).

Project 6   Production of Energy from Waste

Objectives:
(a) To investigate methods for the combustion of low calorific value fuels.

(b) To examine other uses of waste.

Achievements:
1. One project was aimed at the utilisation of low-grade coal shales with a high ash content in fluidised bed combustors. It is believed that the process will be economically viable, though the industrial plant will be very large.

2. Fluidised bed combustion of waste materials was also examined and found to be attractive. A survey of suitable waste materials to serve as feed stock was completed. It was clear that higher combustion temperatures than those usually associated with fluidised bed combustors might be necessary to improve carbon utilisation in the wastes.

3. A high quality granulate suitable for building material has been developed. This is produced from normally undesirable residual material which is presently disposed of at a high cost. A major advantage is that the building material is produced using the residual energy from the waste itself.

Project 7   Industrial Processes

Objectives: This project concentrates not on energy saving by recovery of heat, but by the improvement in the industrial processes themselves, the primary objectives being:
(a) Energy Analyses of industrial processes with the aim of improving their energy efficiency.
(b) Optimisation of selected processes.

Achievements:
1. A number of detailed energy analyses were completed in the food, textile and paper industries which identified possible improvement areas. For example, one study showed that in the refinement of paperstock, considerable quantities of electricity could be saved ($\sim$ 10%) by varying the pulp flow rate and the power to the refiner according to the prevailing conditions and types of paper being produced. In a fine paper mill producing 10K tonnes per year, the payback period is estimated at between 2 and 4 years.

2. A computer program has been developed which will allow companies to select components and determine their optimum arrangement for energy saving in an industrial process. It is applicable to different manufacturing systems.

3. Several studies on combustion and boiler performances identified ways of improving boiler performance, but also indicated clearly the problems that exist in monitoring combustion conditons.

Project 8 Development of Methods of Accumulating Secondary Energy

Objectives:
(a) Investigation of methods of storing low grade heat.
(b) Investigation of advanced secondary batteries for electrical storage.
(c) Investigation of flywheel energy storage.

Achievements:

1. A compilation has been made of materials for latent heat storage of low grade heat over a range of melting points from $-50°C$ to $+130°C$. Detailed experimental investigations have identified about 30 of them as satisfying the quality requirements for practical applications in storage units, and crystallisation and corrosion problems have also been examined.

2. About 30 chemical heat storage systems were investigated and it was shown that for the storage of 1 MWh of energy, a volume of $2m^3$ was required for the gas/solid H_2O/Na_2S system, while $4.5m^3$ was required for the gas/liquid H_2O/H_2SO_4 system.

3. Flywheel storage was examined without any encouraging results; windpower systems were identified as one of the few promising areas of application.

4. A production technique has been developed for ceramic electrolyte tubes (ß — alumina) for sodium-sulphur batteries. These have performed well in practical tests.

5. Lithium-sulphur cells with a capacity of 100 ampere hours have been developed with fused salt electrolytes with energy densities of 80 Wh/kg and a 400 cycle life.

6. A detailed assessment has been made of the likely use of batteries for electric storage in a number of application areas and it was concluded that electric vehicles would continue to provide the largest market during this century.

7. Encouraging preliminary research has been completed in "all solid" batteries with LiAl and TiS_2 electrodes and $LiAl_2O_3$ or polymer electrolytes, and in glassy electrolytes and electrodes. These results suggest that further research work would be profitable in these areas.

Second Programme (27 MioECU)

In the second programme, some of the difficulties of definition encountered in the first programme are recognised, so that the action areas are redefined in terms of the end user as:
— domestic and commercial applications
— industry
— transport

while two other overlapping areas are *separately* defined because of their importance
— energy transformation and transport
— storage of secondary energy.

There is a general view in the allocation of funding that it should be related in some way to the energy saving potential, which is influenced by the level of energy consumption, the energy losses, and the growth rate in the consumption. It is also influenced in part by the availability of fuels. Attention is also paid in the second programme to other factors such as market size, payback time, and opportunities for demonstration and implementation.

In the domestic and commercial sector, R & D will be supported, for example in heating and air conditioning systems including heat pumps, the development of improved insulation materials and processes for the insulation of buildings, the study of buildings as integrated systems, including occupancy factors to gain a better understanding of building performance, the efficiency and effectiveness of energy consuming equipment such as electrical appliances and lighting.

In the industrial sector, emphasis will be placed on the reduction of the specific energy requirement of energy intensive industries, the utilisation of residual heat as an alternative fuel, and the development of energy management techniques.

In the transport sector attention will be paid to the improvement of traffic systems as well as to improving engine efficiency.

Energy transformation and transport will concentrate on efforts to develop more efficient transformation installations and to recover waste heat. The development of technologies to promote the use of coal and low grade fuels will be encouraged insofar as there is no conflict with the terms of the ECSC Treaty.

The energy storage effort will continue on the basis of the progress made in the first programme with the same emphasis and range of areas, and will be co-ordinated with the appropriate parts of the solar and hydrogen programmes.

Heat Wheel applied to a Terylene Oven, (IRD & ICI, United Kingdom).

PRODUCTION AND USE OF HYDROGEN

One common feature of most of the so-called alternative energy sources is that they will produce an increased emphasis on electricity as the principal energy carrier to the user. This creates a number of problems associated with the distribution of fuel to moving equipment at remote areas, and with the storage of energy; at the moment the only available technique for large scale storage of electricity is through pumped hydroelectric schemes.

It is important therefore to identify and develop suitable secondary fuels which can meet both of these requirements. Hydrogen has a number of well known characteristics which make it an attractive contender for this role. It can be easily produced, stored and transported; it can be easily converted into electricity, mechanical energy or heat causing little or no pollution, and it is presently a very important raw material in the chemical, petrochemical and metallurgical industries.

Today, hydrogen is produced commercially either chemically from coal or oil, or electrically by the electrolysis of water. These are long established techniques, but for electrolysis there is considerable room for improvement in the design and performance of the electrolysers in general use. For example, they require the application of a potential of about 2 volts to a cell carrying a current of $2kA/m^2$, which is to be compared with the theoretical value of 1.23 volts. This represents a very large inefficiency which will require an R & D effort if it is to be overcome. There are also some potentially interesting thermochemical approaches to producing hydrogen on a large scale, using reactions that are driven at high temperatures ($\sim 800°C$). These open up the possibility — even if a remote or long term one — that nuclear power stations could be used to produce convenience fuel as well as electricity. Again R & D is essential.

It is expected that, over the long term, the hydrogen market is likely to shift from that of a raw feedstock to that of a large scale energy carrier and contributor to the production of synthetic substitution fuels. This will involve a significant disturbance to the energy supply system and a considerable R & D effort is necessary to effect this movement smoothly. It is worth remembering, however, that the use of hydrogen as a fuel is not entirely new as towns gas produced from coal contained a sizeable proportion of hydrogen ($\sim 50\%$). Thus, the precedent is well established, and R & D effort is needed to improve methods of production and storage of hydrogen and to extend the range of its utilisation. This was the purpose of the first R & D programme, which concentrated most heavily on hydrogen production by electrolysis and by thermochemical methods.

Summary and Achievements of the First Programme (13.2MioECU)

Three study areas were identified for the first programme—
1. Thermochemical production of hydrogen.
2. Electrolytic production of hydrogen.
3. Transportation, storage and utilisation of hydrogen.

Project 1 Thermochemical Production of Hydrogen

Objective: Study and development of this new route to hydrogen production. This project was specifically aimed at expanding and supplementing the work carried out at the Joint Research Centre at Ispra under the "direct action" programme.

Achievements: The method is based on a series of chemical reactions operating in a closed cycle in such a way that the only net inputs to the cycle are water and heat and the only outputs are hydrogen, oxygen and low grade heat.

Assembly of electrolyser for hydrogen production, (SRTI-IFP, France).

Assembly of electrolyser stack for advanced electrolyser type, (CERCEM, France).

Electrolyser for Hydrogen production, (SRTI-IFP, France).

This work was pioneered by the JRC at Ispra and consequently, the indirect action has been closely co-ordinated with the work of the JRC, bringing in specific expertise from universities or industry to further the work at Ispra.

During the development of the programme, many possible routes were discarded for a variety of reasons — materials problems, technological difficulties, difficult reaction steps, etc. The two cycles which appear to be the most attractive of those studied are actually of the 'hybrid' type in which one step involves an electrolytic reaction. These have been designated Mark 11 and Mark 13 and are represented by the equations:

Mark 11 $H_2SO_4 \rightarrow H_2O + \tfrac{1}{2}O_2 + SO_2$ (800°C thermal)

$SO_2 + 2H_2O \rightarrow H_2SO_4 + H_2$ (electrolytic)

Mark 13 $H_2SO_4 \rightarrow H_2O + \tfrac{1}{2}O_2 + SO_2$ (800°C thermal)

$SO_2 + 2H_2O + Br_2 \rightarrow 2HBr + H_2SO_4$ (room temperature)

$2HBr \rightarrow Br_2 + H_2$ (electrolytic)

The indirect action programme originally contributed to the identification of Mark 11 and Mark 13 as the preferred reactions; through screening of possible cycles and possible electrolytic reactions and by contributions to flow sheet synthesis.

Subsequent efforts were directed towards phase separation in the sulphuric acid and SO_2 parts of the cycles, electrolytic oxidation of SO_2, corrosion studies and material selection, sulphuric acid decomposition with the preliminary design of purification unit for removing SO_2 from the output, and with the draft design of a small Mark 11 plant.

Project 2 Electrolytic Production of Hydrogen

Objectives:
(a) Improvement of present technology in the low to medium temperature range (80 — 120°C).
(b) Development of medium temperature electrolysis (150° — 200°C).
(c) Development of high temperature solid electrolyte electrolysis (800 — 1000°C).
(d) Development of new diaphragms or membranes.

Achievements:
1. In the low temperature ranges total cell voltages of between 1.6 and 1.9V have been achieved at current densities of $10 kAm^{-2}$. This corresponds to a voltage of between 1.4 and 1.5V at the lower current densities ($\sim 2 kA/m^2$) of present electrolysers and should be compared with the 1.9 – 2.1V which is common today. Similar promising results have been achieved at medium temperatures. Thus significant improvement has been achieved and shows definite progress towards the ideal and therefore unachievable, value of 1.23V. With this newer cell design a unit cell thickness of 10mm is feasible and this permits a considerable power density as $10 kAm^2$ in a stack 1m high and 1m in diameter would represent a electrical power input of about 1.4MW and a hydrogen production rate of about 300 Nm^3/h^1. Such a design would ensure that the cell stack was no longer the major cost item of the plant.

2. Small prototype (5 — 10kW power input) electrolysers have been built to transfer the results to a larger scale. Their long term operation is also being pursued to check the stability of the components developed in the laboratory.

3. The diaphragms needed for the practical development of these electrolysers have been developed and tested and are now also undergoing long term testing.

Project 3 Hydrogen Use, Storage and Transportation

Objectives: (a) Study of small and medium scale storage of hydrogen.
(b) Examination of specific applications and future markets.
(c) Materials.
(d) Safety.

Achievements: 1. Improvement of hydrogen storage techniques using magnesium hydrides has been obtained, and a new storage method based on cryoabsorbents has been identified. No evaluation of large scale underground storage has been performed but no major difficulties are expected in this area.

Experiment on splitting of SO_3 for thermochemical hydrogen production, (KFA-Julich, W. Germany).

2. Hydrogen applications and market evaluation studies have been carried out in the areas of coal conversion, steel making and other industrial processes, electricity production (in particular, the status of hydrogen fed fuel cells has been quantified) and the likely evolution of the potential hydrogen market has been evaluated to estimate the rate of implementation.

3. The materials programme has shown that no problems are likely to be encountered with existing gas networks, and several steels have been studied and are being optimised for use with hydrogen. Cross tests to arrive at agreed testing methods are being carried out by several contractors.

4. Most of the data necessary for compiling a safety manual for hydrogen operators have been collected, though it is worth emphasising that the safety problems are less with hydrogen than with other fuels in general use.

Second Programme (8 MioECU)

In the second programme, the emphasis has moved somewhat away from thermochemical production and towards electrolytic production and the problems of utilisation, storage and transport. The objectives are to move towards applications, long term testing and the scaling up of promising concepts to a small number of relatively large scale tests of the best achievements by the end of the programme.

Heat pipe development: (a) with wick; (b) with grooves, (UKAEA, Harwell, United Kingdom).

SOLAR ENERGY

The European Community lies between latitudes 35°N and 60°N (excluding Greenland) and, as a consequence, the amount of solar energy available varies greatly over the different regions. For example, the average insolation on a horizontal surface can range (in KwH per sq.m. per day) from 0.21 in Shetland to 2.4 in Sicily in January and from 4.8 in Shetland to 8.0 in Sicily in June. That is, the Sicily summer insolation is some 40 times greater than the Shetland winter value. This large geographical and seasonal variation creates both problems and opportunities for the utilisation of solar energy in the Community.

Solar energy manifests itself at the earth's surface either directly as heat and light, or indirectly as wind and wave power through its effect on the atmosphere. The direct radiant energy can either create heat or can induce photosynthesis and the corresponding growth of green plant material. Thus, if solar energy is viewed as an energy source, any technique which relies on photothermal, photoelectric, photochemical, photoelectrochemical or photobiological conversion, the use of wind or wave energy, or the direct use of biomass materials, which depend ultimately on photosynthesis, can be regarded as a legitimate area of study.

At the beginning of the first programme, there were still a great many questions to be answered at all levels of treatment of the solar energy problem and it was decided that only those areas directly dependent on radiant energy would be examined. That is, wind power and wave power were not included in the EEC programme directly, though biomass utilization was.

In particular, emphasis was placed on the areas in which it seemed that good factual information and experience was lacking, thus, solar energy applications to dwellings, to thermoelectric and photoelectric power production, to the use of biomass, and to the longer term photochemical, photoelectrochemical and photobiological conversion systems were encouraged. Also, the acute shortage of reliable solar data and methods of testing solar collectors were recognised and included as specific areas to be developed.

These efforts were successful in the first programme and achieved their objectives regarding basic levels of knowledge and understanding. In the second programme, the emphasis has shifted to the more practical problem of development and production of prototype systems with the aim of identifying systems problems, proving the viability of solar energy, and encouraging its rapid implementation. One particular feature here is the creation of "pilot plants" which can demonstrate the viability of concepts and techniques on a realistically large scale which is still below what could be called a Demonstration Project. Additionally, wind power was introduced into the second programme.

Summary and Achievements of the First Programme (17.5MioECU)

Six study areas were identified for the first Energy R & D programme:

1. Solar Energy applications to dwellings.
2. Thermomechanical solar power plants.
3. Photovoltaic power generation.
4. Photochemical, photoelectrochemical, photobiological conversion.
5. Energy from Biomass.
6. Solar Radiation Data.

Project 1 Solar Energy Applications to Dwellings

Objectives:

(a) To test and evaluate different collector types and to develop commonly agreed methods for performance rating.

(b) To support the construction of several solar houses in different member countries, and to collect data on existing solar houses with a view to co-ordination of the monitoring of their performance.

(c) Development and validation of models for simulating and predicting the performance of solar buildings.

(d) The investigation of low temperature thermal storage possibilities for solar buildings.

(e) Initiation of research work on solar cooling technology.

Achievements:

Several notable achievements have been made which have enormous consequences for the evaluation of solar systems throughout the Community.

1. The programme for modelling the performance of solar collectors and systems has been largely completed and some valuable analysis programmes have been developed, although not all of the results were conclusive.

2. Standard techniques have been developed and published for the testing of solar collectors through an extensive programme involving 20 laboratories.

3. The programme has been completed for the construction of Solar Pilot Test facilities which allow simulation of the heat load on an actual solar heating system, so that the performance of the solar system rather than just the collector can be assessed under realistic operating conditions.

4. Empirical data has been gathered from 51 solar houses throughout the Community so that the performance of these houses can be more fully appraised. The results have been published in *Solar Houses in Europe* (see appendix 1).

These four achievements obviously contribute mutually to a better understanding of solar systems and to the provision of data and facilities of the better design of solar buildings. One of the important features is that they have involved extensive collaboration between research groups in the various member states, including the involvement of the Joint Research Centre, and have achieved a standardisation of the techniques to be used throughout the Community.

5. In energy storage, the achievements are negative rather that positive. That is, the results were not entirely encouraging and serve to reduce the attractiveness of latent heat storage — at least in the short term. For example, latent heat storage with melting points of about 35°C has proved to produce smaller reductions in storage volume than those that were hoped for. Further, the expected improvement in efficiency brought about by the large latent heat transfer at a relatively low temperature, did not materialise in practice. Thus, the most promising short term thermal store is apparently a well stratified water tank. Long term options are still open and include sensible and latent heat, and chemical storage possibilities.

6. Solar cooling has made some advances — mainly in the development of high performance solar collectors based on sealed and evacuated units. These are essential for providing the high temperatures needed for operation of the system.

The search for effective absorbent-absorber pairs continues, but progress is inevitably slow in such a difficult and time consuming area. One hopeful achievement, however, is the design of a free piston Rankine cycle machine which has now operated under laboratory conditions and which is producing encouraging results.

Project 2   Thermomechanical Solar Power Plants

Objectives: (a) To investigate the viability of large scale electricity production by thermomechanical means using solar power as the heat source.

Achievements: The major achievement is the development, design and construction, at Adrano in Sicily, of the world's first large scale helioelectric power plant to be connected to an existing utility grid. The experimental plant is called Eurelios and has an electrical capacity of 1MW. It has a thermal capacity of 4.8MW and a total mirror surface of 6216m^2. This made up of 252 heliostats, of two different designs. The receiver (furnace) is at a tower height of 55m and comprises a specially designed once-through cavity type boiler producing steam at 512°C, 64at which is fed to the turbogenerator. The system also includes a molten salt + water heat storage vessel with a storage capacity of 1/2 hr. The power plant is built by an industrial consortium consisting of ENEL/ANSALDO, Italy, CETHEL, France and MBB, Germany.

Project 3   Photovoltaic Power Generation

Objectives: (a) i) Assessment of promising alternative solar cells.
ii) Improvement of silicon and cadmium sulphide cells.

(b) Feasibility study on new concepts such as organic dye/semiconductor combinations, organic solar cells, spectral shift devices or semi-conductor/electrolyte systems.

(c) Development of new techniques for the preparation of semiconductor (solar) grade silicon powder from sand. Support for the industrial application of these new processes with special emphasis on energy-saving methods.

(d) Development of continuous production processes for thin silicon sheets, with special emphasis on low energy consumption processes.

(e) Support for industrial scale efforts to reduce costs by improving and automating solar panel production, (including light concentrators and pilot plants).

Achievements: 1. Through the use of automation techniques research efforts have resulted in a reduction in the cost of Si-photovoltaic panels by a factor of 3 to 4 and it seems likely that even without a technological breakthrough a further factor of 2 will be achieved within the next few years.

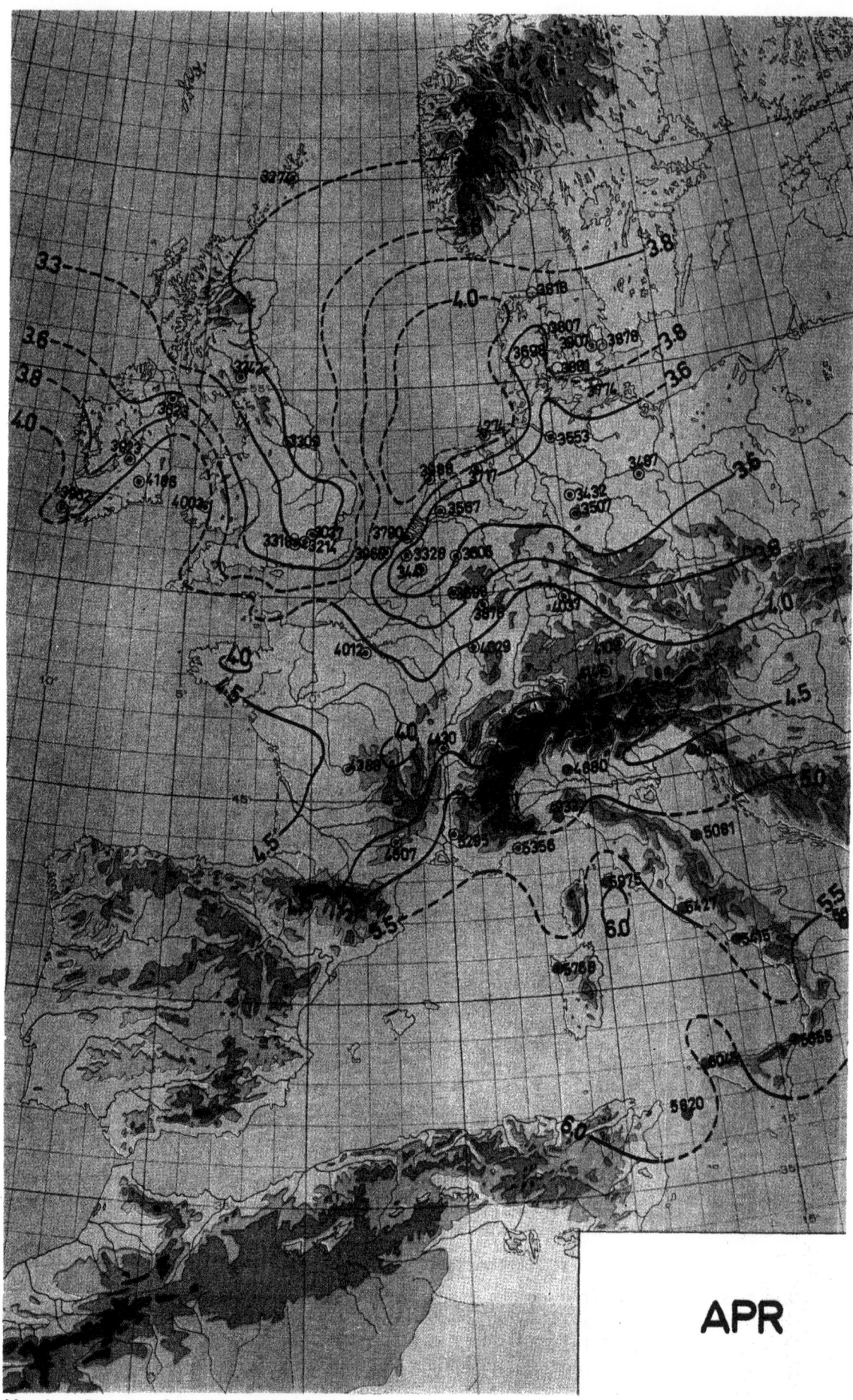

Map from European Solar Atlas, (CEC, Brussels).

2. The use of light concentration (focussing) techniques can reduce the cost of photovoltaic electricity production by a factor of 1.5 in Northern Europe and 2.5 in Southern Europe. It has been possible to increase the usable focussing level through the introduction of active cooling which results in both a more economic system, and a dissipated heat component which can also be used to good effect. However, it still seems that focussing systems will be uneconomic because of the cost of the structure.

3. Because of these two developments the position now is that photovoltaic power plants without storage can be competitive with diesel generators under certain circumstances. The most interesting application for such systems is for large scale water pumping for irrigation purposes.

4. Solar cell module production in Europe has been encouraged to the extent that it is now over 1MW per year.

5. Seven small (5kW or less) photovoltaic systems have been built (2 with concentration, 5 without) to provide system design and operational data for a variety of applications and environments.

6. None of the alternative cell types has yet advanced sufficiently to pose a serious challenge to single crystal silicon although some progress has been made, particularly with cadmium sulphide and amorphous silicon.

Heliostat for Eurelios solar furnace (see page 2).

7. Under the ægis of the Joint Research Centre at Ispra a specification has been produced by a small international working group of experts which lays down standard procedures for photovoltaic performance measurements. This should provide a common basis for performance rating which will remove the discrepancies which have arisen in the past.

8. Design studies have been completed on large (0.5 — 1MW) photovoltaic power systems with storage and grid connection. This programme was included to prepare for the larger scale production units foreseen for later phases.

Project 4 Photoelectrochemical, Photochemical and Photobiological Processes

Objectives:
(a) Understanding of photoconversion processes.
(b) Improvement of hydrogen production by living cells construction of synthetic systems based on a knowledge of photosysthesis.

Achievements: These objectives were approached by a programme of fundamental research on endothermic photochemical, photobiological and photoelectrochemical processes related to solar energy utilisation.

1. A beginning has been made towards understanding how the thylakoid membrane functions as an energy converter. For example, it is known that the first reduced compound to be identified in the membrane is an iron sulphur protein which contains more energy than hydrogen reduction. If the operation of this chloroplast membrane in converting sunlight into chemical energy can be fully understood, then it should ultimately be possible to design analogous artificial systems.

2. The possibility of hydrogen production from industrial wastes using solar energy as the energy source has been opened up by studies on cyanobacteria (blue-green algae) and green algae, which are capable of producing hydrogen and oxygen by decomposing water, and other photosynthetic bacteria which can produce large quantities of hydrogen or ammonia with light and simple organic or inorganic substrates. Other related studies are aimed at understanding the processes involved.

3. A device has been constructed which uses the chloroplast membrane to generate a current by a photogalvanic cell. It is important to appreciate the scale so as to fully understand how far this system is from engineering reality. An open current potential of 220mV was generated in a cell with platinum electrodes. This resulted in a current of $800 \mu A$ and a current density of $16 \mu A/cm^2$. The cell can be modified to produce H_2 and O_2 instead of electricity.

Project 5 Energy from Biomass

Objectives:
(a) Selection and promotion of the most suitable energy crops for various regions of Europe.
(b) Adaptation and appraisal of the technologies for harvesting, drying, transporting and utilising energy crops.

Achievements:
1. The most promising routes to the production of energy from biomass have been identified and it is believed that biomass energy sources could contribute up to 10% of EEC energy demand by the year 2000.

2. The amount of the available resources has been estimated (cf. *Energy from Biomass in Europe*, W. Palz and P. Chartier, Applied Science Publishers, London 1980).

3. A number of small devices to study the chemical and other operating conditions of anaerobic methane digesters have been built.

4. Units for direct gasification at temperatures about 1000°C have been built for studying this possibility. In particular a gasification plant of about 1MW output has been constructed.

5. A number of successful studies of catch crops, sylviculture and algae culture have been completed. For example, short rotation forestry on various types of peat soil conditions has been carried out in Ireland so successfully that it has now become an EEC Demonstration project.

Project 6 Solar Radiation Data

Objectives:
(a) Collaboration with the national data centres.
(b) Production of reference years for simulation of solar components and systems.
(c) Preparation and publication of Atlases of solar radiation over the Community.
(d) Research on new measuring equipment.

Achievements:
1. Towards the first objective, very good agreement was achieved between the national data centres in the calibration and comparison of pyrheliometers. Problems were identified with pyranometers and the comparison is being extended in the second programme to cover these.

1kW photovoltaic electricity generator, at Florence (Montedison/ANSALDO/GALILEO/SGS-ATES, Italy).

2. A technique was established for producing a test reference year data set including at least dry bulb temperature, global, diffuse and direct normal radiation, wind velocity, with humidity, cloud cover, cloud types and wind direction being included if possible. A test reference year is a collection of as much as possible of the above weather data, arranged as sets of simultaneous readings taken at hourly intervals (8760 data sets) and chosen so that each month in the data set is typical of the location.

3. An Atlas of the European Community has been compiled showing the insolation on a horizontal surface in all countries. It was compiled from data taken at 56 selected stations throughout the Community and daily totals were made of global radiation and sunshine duration during the period 1966-1975. These were converted to a common format and subjected to quality control procedures. Tables of monthly totals are presented and also maps showing the isolines of global radiation. The atlas is intended to be a working manual suitable for architects and solar engineers, and has proved to be very popular.

4. It was shown that while hourly data was adequate for most purposes, some projects do require data to be sampled at shorter intervals (six to ten minutes) and measurements of this kind have been undertaken together with their statistical evaluation.

5. The statistical study of hourly values of global radiation on a horizontal surface has made it possible to rationalise the concept of cumulative frequency curves and their use for predicting the performance of solar thermal collectors.

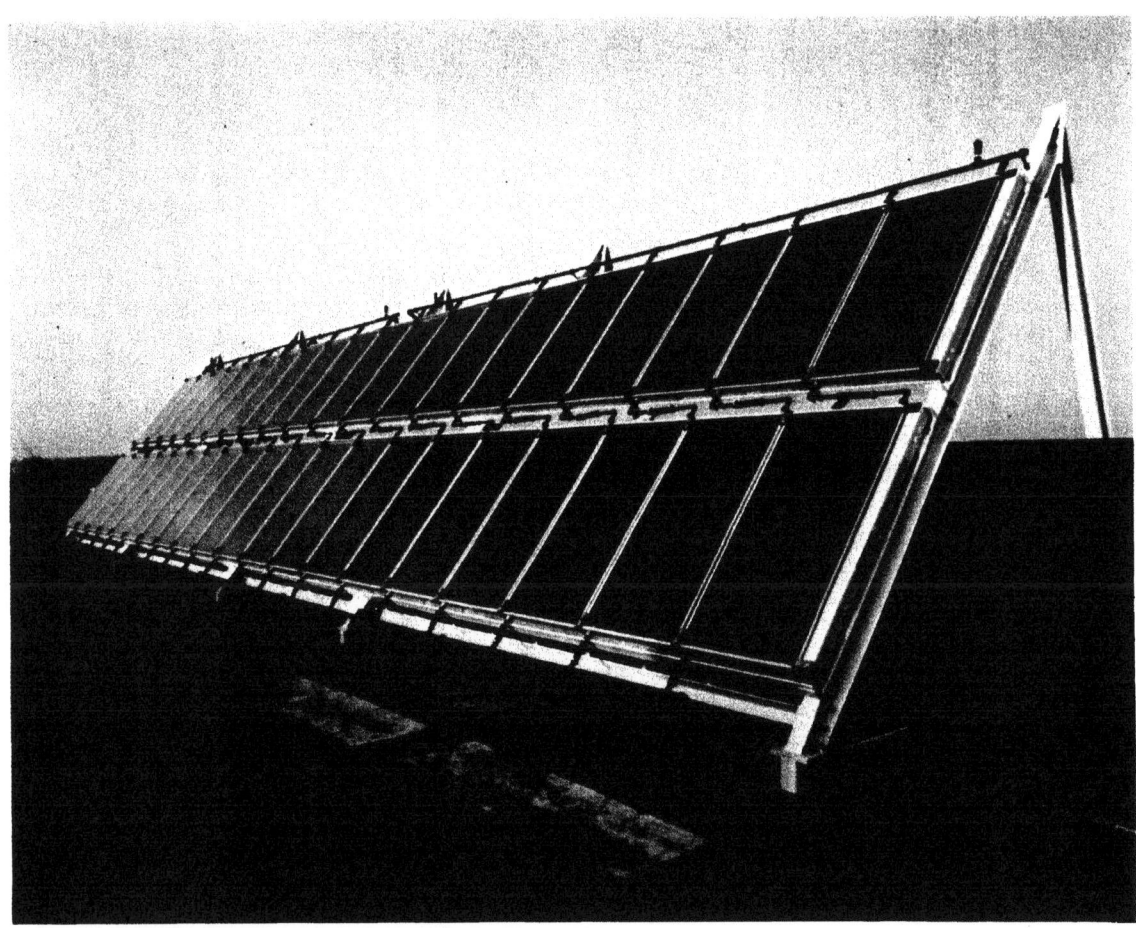

Solar pilot test facility, (Bracknell, United Kingdom).

6. Methods have been developed for using satellite images for the prediction of solar radiation levels.

7. Progress has been made towards the extension of the measurement network.

Second Programme (46 MioECU)

In the second programme, the basic objectives of the First Programme were retained, that is the same six projects are present, but two more: *Wind Energy* and *Solar Energy in Agriculture and Industry*, were added. Thus, there is an apparent continuation of activity. However, there is a significant change of emphasis in the current activities; instead of the earlier exploratory work, a large emphasis is now on the development and construction of prototype systems with the aims of identifying system problem areas, increasing the credibility of solar energy, and encouraging its rapid implementation. These prototypes must now, as a rule, have a certain minimum size if they are to be credible as future large scale energy sources and are to allow testing of not only the solar components, but all of the other elements required for implementation of a complete system. European industry will be encouraged to specialise and to co-operate effectively in the solar energy sector.

Specifically the objectives are:

1. To continue the investigation of cost effective heat storage techniques, to develop and test solar heating system models, to test collectors and systems, to further investigate solar cooling methods and to integrate solar concepts into building design.

2. To complete and operate the 1MW Eurelios helio-electric plant.

3. To continue work on solar collector arrays so as to reduce costs and increase life-time, to design and develop a family of photovoltaic power systems, throughout the Community, with a total power of 1.3 MW, to examine operation and application in various European climates.

4. To continue the basic research aimed at obtaining a better understanding of the mechanisms of photochemical, photoelectrochemical and photobiological processes. The fundamental objectives are the same as in the first programme.

5. To initiate a small number of pilot projects on synfuel production — particularly methanol from wood, but some effort is foreseen on the production of ethanol, biogas and some schemes for biomass growing.

6. Further development of solar atlases and data books of solar radiation in the European Community for simple design methods, extension of data banks, development of measuring instruments.

7. In wind energy, research is envisaged on site evaluation and general potential estimation in the European Community, investigation of wind generators for offshore operation, testing under various climatic conditions, combination of wind generators with photovoltaic systems, biomass production plants, etc.

8. Assessment of energy needs and potential solar applications in agriculture and industry, development of systems models, and monitoring of advanced systems such as greenhouses, drying systems and steam generators.

GEOTHERMAL ENERGY

Geothermal energy has been confirmed as being potentially a significant contributor to the Community's supply of energy from indigenous resources. However, its expected contribution is limited by three factors:

1. Most of Europe's geothermal resources are available at a low temperature.
2. It is necessary to have water associated with the heat source to permit heat transfer to the surface.
3. Deep drilling is expensive so that the exploitation of geothermal energy is economic only at certain privileged sites where there are geothermal reservoirs at reasonable depths with the additional constraint that there must be a suitable market for the heat.

There is a possibility that the second limitation might be removed if it were possible to fracture hot dry rocks and inject water from the surface. If this could be done, the potential for geothermal energy application in the Community would be very much larger, but much work remains to be done after some very encouraging results in the first phase of the programme. The development of hot dry rocks as a useful geothermal source must be regarded as a long term aim.

In the Community, the utilisation of geothermal energy has been limited in the past to two countries: Italy and France. In Italy, steam fields have been exploited for the production of 350MW of electricity (423MW installed capacity), while in France, hot water sources in the Paris and Aquitaine Basin have been developed for the production of 26MW of heat for space heating.

The aims of the Community R & D effort were to build on this base, to determine the possibilities of future wider application, and to improve exploration and utilisation techniques.

Summary and Achievements of the First Programme (13 MioECU)

Five projects were identified for the first R & D programme.

1. The acquisition and collection of existing and new geothermal data;
2. The improvement of exploration methods;
3. The exploitation of hot water sources (low enthalpy);
4. The exploitation of steam sources (high enthalpy);
5. The development of techniques for use with the exploitation of hot dry rocks.

Project 1 Acquisition and Collection of Existing and New Geothermal Data

Objectives: (a) To accumulate and compare all possible appropriate information in the nine European Community Member States (subsurface temperature distribution, temperature gradients, heat flows, etc.)
(b) To identify possible reservoir areas and to assess the energy content and possible yeild of selected reservoirs.

Achievements: Some of the achievements — both positive and negative — of the first programme can be illustrated by three examples.

1. The production of a comprehensive Atlas of subsurface temperatures throughout the European Community. This initiative stimulated national projects for collecting information on subsurface tempera-

ture gradients and heat flow, and is an extremely valuable working tool, relating temperature characteristics to geological factors. It must be admitted, however, that the quality of the data is often variable as it is inherited from oil exploration efforts which are not aimed at studying thermal conduction in geological formations. Notwithstanding these limitations, the Atlas provides the most comprehensive data base available to the European geothermal engineer.

2. Downhole instrumentation has made significant advances within the Community programme. This is a critical area as without adequate and accurate information on geothermal flows in the borehole itself, no useful picture can ever be built up about the nature of a geothermal field.

    Two particular projects deserve special mention. First is a probe based on a completely original measuring technique involving microwelding techniques with thermopiles derived from cardiac pacemakers. This is an important technological advance. It has successfully completed its initial test and development phase and is now undergoing extensive in-situ assessment.

    The other major development is a composite electronic logging instrument which can measure two-phase flow rates for a combination of water and steam under operating conditions at downhole temperatures and pressures of 250°C and 250 bars. This is a robust and versatile electronic instrument which will be of enormous benefit in steam reservoirs whose monitoring has previously been dependent on less satisfactory mechanical devices. It is currently being developed to 400 bars and 300/350°C.

3. Since oil and mining operations are continually taking place on a world-wide basis, it seemed sensible to investigate the possibility of acquiring geothermal information at the same time. Unfortunately, it has proven to be difficult to modify such a mining or petroleum research programme and additionally where European Community research money has been injected, the quality of the data obtained has often represented an insufficient return on investment. Nonetheless, the experience has been valuable and has highlighted the need for modified exploration and drilling programmes. It has also provided part of the basis for future detailed reservoir surveys in the United Kingdom, Northern Italy, the Netherlands, North Germany and Ireland.

Project 2 Prospecting Methodology

Objectives: To improve the indirect assessment of geothermal indices from subsurface and surface measurements.

Achievements: Prospecting methodology was approached by adopting a range of geophysical and geochemical techniques and by the undertaking of integrated geophysical surveys.
It is probably fair to say that efforts to establish one or another technique as an optimum indicator were not successful, but instead, it has appeared that all possible techniques should be applied and used in conjunction with each other. Certainly, individual successes can be associated with some projects such as the appearance in electromagnetic soundings of a clear correlation between deep seated conductive layers and previously known hot fluid circulation in the Phlegraen fields (Naples — Pozzuoli area), or the development of a

Cyclone separator at well site MO2 at Mofete in the Phlegraen active volcanic high enthalpy geothermal field, (AGIP, Italy).

new Na/Li geothermometer applicable to fresh water or brine. This instrument is presently being checked on fluids produced recently in the Phlegraen fields at 310°C extrapolated temperature.

More important, however, is the success of detailed integrated exploration surveys in which several geophysical and geochemical techniques were combined. For example, at Mont Dore, in the Massif Central in France, a combination of electric and electromagnetic soundings, high and low frequency magnetotelluric measurements, gravity and shallow refraction seismic and deep seismological measurements made it possible to identify a caldera which was not evident from topography. Field trips and dating estimates allowed the reconstruction of the volcanic scenario, and particularly the past caldera events which control hydrothermal convection processes. Geothermometers (based on SiO_2 Na-K and Ca) assessed spring water equilibrium temperatures of 150-160°C with fair reliability to the North West of the site in granitic basement rocks. These trends were supported by hydrogeological considerations, regional heat flow patterns and electromagnetic soundings, evidencing deep conductive horizons in the basement. The result of these studies is that the existence of a high enthalpy (steam) reservoir at Mont Dore should be regarded as unlikely but a scientific wild cat (exploratory) well is projected to conclude this multidisciplinary approach, and check the validity of the conceptual picture which has been built up.

Project 3 Utilisation of Hot Water Sources (Low Enthalpy)

Objectives: To cover all theoretical and practical aspects of low enthalpy geothermal energy exploitation with special reference to urban heating and industrial and agricultural applications.

Achievements: Because of the fact that low enthalpy sources are the most widely distributed throughout the member states, this area is more extensively covered in the programme than steam sources. The project involved theoretical modelling of geothermal reservoir behaviour, full scale experimental verification of these models, and investigation of exploitation and reservoir management problems. It should be noted that with some difficulties petroleum and ground water technologies can be applied to low enthalpy problems since all three share similar sedimentary reservoir conditions.
There have been a number of specific achievements within the project:

1. Two contracts dealt with the design and spacing of the pair of wells used for extraction of the geothermal fluid and its re-injection after heat removal (the couplet). The objective was to maximise the lifetime of the reservoir. A number of factors such as thermal conductivity of the cap rock and the aquifer, or the effects of viscosity changes with temperature and pressure, were investigated and it transpired that it is possible to improve the lifetime of the reservoir though only at the expense of a reduction in the heat output. The effects of interaction (or short circuiting) between the inlet and outlet sides of the couplet have been analysed and it has been shown that suitable arrangements can increase reservoir lifetime through production of an adequate waterdrive. The influence of reservoir inhomogeneities has been investigated and efforts are now to be placed on the proper experimental characterisation of reservoir structure.

2. On the utilisation side, one important result has been derived concerning the superficially attractive combination of low temperature geothermal sources with heat pumps. At Creil in France, this has been examined and it proved possible to raise the heat supplied from 40% to 64% of the demand. Unfortunately, the economics look somewhat discouraging with a payback period of 30 years assuming a 9% discount rate. Thus a cautionary result is obtained which implies that the areas of combined use of heat pumps and geothermal heat must be carefully studied before any large scale development is made.

3. Much more hopeful is the development of a low temperature convector system suitable for retro-fit exploitation in existing buildings. This allows the use to 50°C water supply temperatures for space heating and, after a very successful development period, it is now undergoing extensive field trials.

Project 4   Utilisation of Steam Sources

Objectives: Improvement of background knowledge on the exploitation of high enthalpy fields, reservoir management and associated technologies.

Achievements:

1. Mud and cement mixtures have been tested for high temperature and pressure application. In particular, an API cement has been developed for down hole application up to 27°C and 500 bars.

2. The industrial prototype down hole probe referred to earlier was developed and underwent successful field tests with accuracies better that 1% in pressure and 0.25% in temperature.

3. Stimulation of dry wells at Larderello in Italy by using cyclic water injectors to stimulate rock fracturing by a process similar to fatigue yielded controversial results. Increased permeability and injectivity were achieved but the mechanism is in doubt. It could well be that the improvements are due to bore hole flushing rather that to induced rock fracturing.

4. Fundamental laboratory studies were carried out on heat and mass transfer in systems involving in situ vaporisation in conjunction with geothermal production and re-injection. One interesting result relates to the dynamics of the vaporisation front.

Project 5   Hot Dry Rocks

Objectives   Feasibility appraisal of heat recovery and power generation from hot dry rocks.

The exploitation of hot dry rocks must be regarded as only a long term possibility. Certainly, if present work is successful, and if the system can be made economically viable, then the potential in the Community is large. The background lies in the Los Alamos concept which proposed the circulation of water through a deep loop of fractured quasi-impervious crystalline rock. This acts as a heat exchanger for recovery of heat from basement rocks at depths of five to six kilometres.

Experiments at a depth of 3000m in New Mexico (Valles Caldera) showed the many problems that can occur. For example, with a penny shaped fracture zone, an area of 5km^2 would be needed to achieve a 50MW output, and this was not achievable in a single crack. A multiple path was required to connect two wells. There were also many other problems related to hydro frac techniques and the effects of channelling — which would introduce preferential paths within the crack — and so forth.

The approach in the European Community project was to tackle the problem of crystalline rocks at shallow depths, to link two wells by a single path and then by multiple paths, to introduce reliable crack monitoring techniques, to circulate a fluid and evaluate loop impedance and heat recovery efficiency, and finally to assess these shallow depth experiments before attempting experiments at greater depth.

Achievements:

1. The Camborne School of Mines in Cornwall (UK) has successfully linked together existing cracks which have long been observed in quarries and in mine galleries in Cornish granites. The linking process combined the use of a low charge explosion fired at bottom hole (300m) to create a radial array of cracks with lengths equal to up to ten times the bore hole radius. Development of these cracks was then extended by hydraulic pressurisation with the aim of connecting existing fracturation patterns and bore holes.

   This was successful, and three wells were linked by at least three fracture paths thought actually to be re-opened joints. The cracks remain open and self-propped. Additionally, the fractures are vertical. Quantification of the field results is still under way.

2. The French Oil Institute has undertaken experiments in high temperature (200°C) and pressure (200 bars) leaching of crystalline rocks from core samples of various origins. Hydro-alcoholic solutions of sodium hydroxide seem to be the most effective leaching agents for material dissolution and precipitation limitation. This work is important, since if large depth fracturation proves possible, chemical leaching would be a promising means of improving both the fracture conductivity and the self propping.

Second Programme (18 MioECU)

Building on the experience of the first programme, the second is developing in a slightly revised form. There is continuing emphasis on assessing as accurately as possible the total resources and recoverable reserves of the Community states, and this is being pursued on what is essentially a regional basis. Areas within the various member states which require more detailed surveys, have been identified and these are being pursued — the ultimate aim being the drilling of an exploration hole. Research is continuing on detailed exploration, feasibility studies and the economic aspects of geothermal energy, though basic research will continue to ensure a proper scientific backing for these projects.

In view of this new approach, the number of study areas has been reduced to four with a slightly different classification:

— Integrated geological, geophysical and geochemical investigations in selected areas.
— Sub-surface problems of natural hydrothermal resources.
— Surface problems of natural hydrothermal resources.
— Hot dry rocks.

ENERGY SYSTEMS ANALYSIS AND STRATEGY STUDIES

The supply of energy to a diverse economic system such as the European Community is a complex business, and requires many decisions to be taken at both a practical and a political level. To ensure adequate energy supplies for the future, in face of the large number of uncertainties associated with fuel available, consumer behaviour and the politics of supply, is even more difficult and decisions taken now will have far reaching consequences.

In order to decide intelligently between the possible options it is essential to develop a practical feeling for this very complex system so that the consequences of given hypothetical decisions can be evaluated.

This is the business of energy systems analysis as we understand it. A very large number of factors must be included when exploring future energy demand and supply possibilities; their inter-relationships are very complex and, further, they are time-dependent. Taken together with the fact that periods of crisis may aggravate or exaggerate some of the factors, it becomes apparent that mathematical models are an essential aid in describing the evolution of the system and the use of computers is necessary for processing the enormous amount of data involved. In principle, such models should include an allowance for physical, technical, economic, sociological and psychological factors, since all play a part in determining the final behaviour of the system. In practice, only part of each group of factors can be specifically allowed for, and other parts have to be approximated or ignored.

The European Community Energy systems analysis sub-programme was formulated in order to provide a scientific basis for the handling of middle and longterm energy problems at Community level. Essentially, the Community is treated as a group of national economies with a strong measure of economic interaction between them, rather than either an economic whole or a set of independent entities. Thus, the Community model is viewed as a multi-national model built up from the individual national models, but with a strong interaction term. Linkage with the overall world environment is seen as being looser, though the Community represents a geographical and economic unit of sufficient size to make such world comparison significant.

While the primary objective of the sub-programme was to try to gain a better understanding of the energy supply and demand system of the European Community, two other opportunities are also available. If successful, it can provide an aid towards formulating long term energy policy, and it can help in the identification of priority R & D areas and the development of R & D strategy in the energy field.

Given these objectives, the decision was taken to undertake the formulation of an European Community energy system model at a time when such modelling was really in its infancy. Consequently, it was realised that expertise in this area in the Community was rather limited and unevenly distributed, and it was felt that that the establishment of a relatively small closely knit group of experts throughout Europe should be actively encouraged. This group was to consist of smaller national sub-groups together with a contribution from the Commission who would collaborate in developing and validating the computer programmes and the overall system. The necessary work would be divided between the various sub-groups, with close co-ordination by the ACPM. It was also decided to build upon national experience, wherever possible.

The modelling of an energy system has two sides, supply and demand. Ideally, the whole system should be treated as one, but it is a reasonable approximation to assume that supply and demand are separate and that the supply can always be made to match the demand rather than the reverse. For the modelling of the supply side, a technical rather than economic approach has been used which gives a realistic view of the energy supply system. That is, essentially a traditional systems analysis approach has been adopted which introduces basically technological parameters for growth and new supply possibilities. The only economic parameter to appear here is cost, which will determine the attractiveness of a given technology.

One set of definitions must be made clear here. In energy analysis it is important to distinguish very clearly between, for example, the fuel burnt by a power station, the electricity (or electricity and usable heat) sent out from it, and the useful contribution that is eventually obtained by the consumer. Thus, one can define:

Algae cultivation at Calabria, (GSUF, W. Germany).

Primary Energy: the energy content of the original fuel or other energy source.
Final Energy: the energy content of the fuel or other energy form after any conversion processes such as oil refining or electricity production.
Useful Energy: the energy content of the useful application of the energy. This may be in term of heat delivered, or litres of petrol per 100km in the transport sector, or tonnes of cement in the energy intensive industries sector.

Having made these definitions, conversion factors can be identified which relate quantities in the different categories. This is clearly the role that is played by energy supply models which identify the technological character of the supply, conversion and distribution industries and create the Primary Energy requirements from the Useful and Final Energy demand pattern.

The other side of energy modelling is *demand*. This can be handled either by econometric methods with a maximum time horizon of about ten years, or it can be handled by scenario methods which are less limited in their time horizon. Econometric models basically use the experience of history and recent trends to extrapolate from the present position. For these reasons they cannot be relied upon to a degree greater than the likely continuity of the established trends. This is the reason why they have only short-to-medium term validity. In the scenario methods assumptions are made regarding the conditions, attitudes, policies, technologies, etc. prevailing at a given time, and these are then translated into Useful Energy and Final Energy requirements. While it is obvious that these models are less affected by time horizon limitations, they are equally obviously extremely sensitive to the quality of the assumptions made. Both approaches are followed in the European Community programme.

Summary and Achievements of the First Programme (3.88 MioECU)

Objective: To better understand the overall energy supply and demand system of the Community, and so to contribute to the formulation of more rational Energy and Energy R, D & D policies.

The approach to this objective was to establish three specific tasks:
1. To carry out an Analysis of National and Community Energy Systems.
2. To develop adequate instruments (or to adapt existing instruments) e.g. mathematical models to describe these systems.
3. To prove that the instruments *are* adequate — that is, to undertake experimental case studies.

Achievements:
1. An active European group of experts has now been established.

2. A linear programming energy flow model EFOM has been constructed which develops, for an individual nation, the optimal energy supply infrastructure to permit the energy needs to be satisfied under various criteria, e.g. minimum cost. These needs will include factors such as transport, heat supply and energy intensive products such as steel, cement, paper and the like. The model allows the demand for such services (Useful Energy) to be translated first into a Final Energy requirement (in which material products such as cement and steel are translated into their energy equivalents and requirements such as tonnes of coal etc.), and then into a Primary Energy demand that must be met by supplying fuel.
This transformation can be effected under the influence of a number of assumptions and the optimal supply mix developed for each assumption. The significance of this model is that, if the energy based requirements of a community are specified (even in tonnes of cement, etc.), then the optimal fuel mix to meet these requirements can be determined under whatever supply and conversion constraints are assumed.

Cambourne School of Mines quarry site showing extensive jointing of the granite.

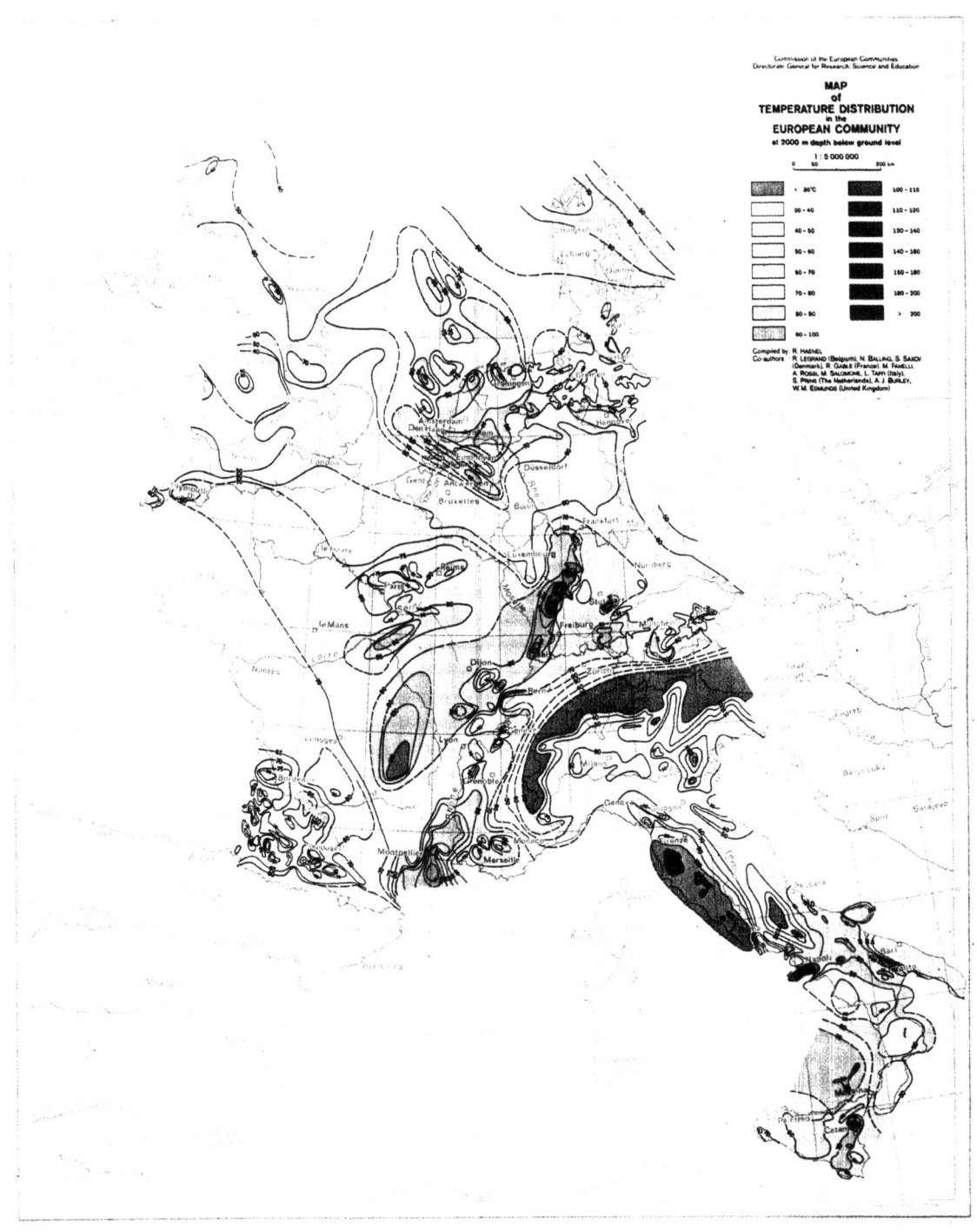

Map from Atlas of Subsurface Temperatures in the European Community, (CEC, Brussels).

3. MEDEE — scenario based model for predicting final energy demand from assumed social, technological and other factors has been developed and tested. It simulates the evolution of various factors that determine energy demand (these are established historically) and calculates the corresponding useful or final energy demand every five years. The programmes are complete and have been applied to most European Community countries.

4. Econometric models for short and medium term energy demand projection have been developed and tested as three separate components:
EURECA — a macroeconomic model modifying the well-known Cobb-Douglas relation between capital, labour and GNP to include a total energy term. This has been developed as a series of national models together with a linkage block which allows for the interaction between the economies of the member states;
EXPLOR 2 — This takes the overall GNP, estimated from EURECA, and determines the size of the different sectoral developments of the economy on the basis of historical experience. It then uses this as one of the inputs to an iterative routine which ultimately produces the outputs (in monetary terms) of the production sectors of the economy.
EDM — which takes the economic production sector intensities and converts them into final energy demand through the use of correlation equations determined from historical values and extrapolated by some appropriate functions. In fact, two different versions of EDM are used, with two different behavioural functions of which one is a single equation and the other a set of simultaneous equations.

The various econometric models developed in this sub-programme work, but some difficulties because of discontinuities in historical trends, have been identified through the application of the experimental case studies. In particular, some of the equations have had to be re-estimated to improve their agreement with recent data. This is a difficult process for smaller countries because the economic statistics are not all available in the same form.

Second Programme — (6 MioECU)

The objectives of the second programme were to achieve full operability of the models, to reconcile the above mentioned scattering in data and equations and to undertake research on new modelling. Both of these are underway and the main results are that on the demand side the equation revision and the adaptation of the data have been achieved, while on the supply side the data harmonisation between countries has been completed. Thus, the way is now clear to complete the multi-national model runs and to finish the experimental case studies.

In the new research areas, work has begun on the development of a new tool for better assessing the relationships between the medium term development and specific energy needs of the main sectors of the European economies, on first exploratory work towards the determination of priorities in the field of energy R & D, and on co-operative efforts in World Energy modelling studies.

CONCLUSION

It is apparent from the foregoing sections that the First Energy R & D Programme has had a large number of successes over a very wide range of topics in the evolution of new techniques and equipment, the advancement of our understanding of energy systems and the processes involved, and in the establishment of definitive objectives for future R & D efforts. Of particular note is the way in which a true European R & D climate has been created with close collaboration between laboratories, both academic and commercial, throughout the member states. It is also significant that this collaboration has developed to an extent far exceeding the actual contracts and financial commitments of the Community R & D Programmes. In effect, these programmes and the contacts induced by them have had the effect of stimulating collaborative work in areas outside the actual programmes. There has also been an element of indirect co-ordination of research throughout the Community through the publicity given to the programmes and its consequent effect on the various national programmes.

The success of the First Programme is reflected in the reports of the Energy Research Evaluation Team (ERET) which was employed by the Commission to examine and evaluate the results of the Energy R & D Programme. Its report on the Energy Conservation and Solar Energy sub-programmes (Report EUR6902EN, 1980) has already been published and it seems appropriate to finish this paper with an extract from the conclusion to that document:

> "The Energy Conservation and Solar Energy Sub-programmes of the first Energy R & D programme can be considered as very successful, account taken of their novelty and of the framework in which they developed. Furthermore, a good number of the recommendations identified by ERET have been or are being independently introduced by the Commission.
>
> ....
>
> ....
>
> It is too early to draw a balance as to their actual or potential impact on the energy situation of Europe: the first results are flowing in and even when these refer to relatively short term applications the likelihood and speed of their penetration into the market still has to be demonstrated. Other results, although very promising, refer to longer term perspectives that will take several years before developing into practical schemes. However, specific examples have been identified where the results can definitely be expected to produce returns in terms of energy conservation or use of renewable sources.
>
> European co-operation is an important aspect of the programme, and, in ERET's view, an objective itself. Some good examples of such co-operation have been identified in both the Energy Conservation and Solar Energy sub-programmes. The regular meetings among contractors working on related subjects have proven to be a very effective way to stimulate international co-operation and co-ordination of efforts. This and other means should enhance the international aspects of the energy programme in the future.
>
> Collaboration of industries, research institutes and universities has been achieved in several instances and should be further pursued as it is also important for the practical diffusion and implementation of results."

The second ERET Report, dealing with the other three sub-programmes, has been completed and is in the process of publication. It is already known that its observations are equally positive and constructive.

APPENDIX 1

Conferences and Publications Arising Out of the First Energy R & D Programme

	Conferences	Proceedings Published by:
1.	First Photovoltaic Solar Energy Conference, Luxembourg, Sept. 1977	D. Reidel, Dordrecht (1978) EUR 5913 EN ISBN 90-277-0884 4
2.	Seminar on Geothermal Energy Brussels, Dec. 1977	EEC, DG XIII, Luxembourg (1978) EUR 5920 EN
3.	Workshop on Concentrators for Solar Energy Applications, Louvain, Belgium, Sept 1978	
4.	Hydrogen as an Energy Vector Brussels, October 1978	EEC, DG XIII, Luxembourg (1978) EUR 6085 EN
5.	Solar Energy for Development Varese, March 1979 Technique et Documentation, Paris, 1980	Martinus Nijhoff, The Hague (1979) EUR 6377 EN ISBN 90-247-2239 EUR 6377 FR ISBN 2-85206-047-4
6.	Second E.C. Photovoltaic Solar Energy Conference, Berlin, April 1979	D. Reidel, Dordrecht (1979) EUR 6376 EN ISBN 90-277-1021-X
7.	New Ways of Saving Energy Brussels, October 1979	D. Reidel, Dordrecht (1980) EUR 6660 EN ISBN 90-277-1078-3
8.	Energy Systems Analysis Dublin, October 1979	D. Reidel, Dordrecht (1980) EUR 6763 EN ISBN 90-277-1111-9
9.	Hydrogen as an Energy Vector Brussels, February 1980	D. Reidel, Dordrecht (1980) EUR 6783 EN ISBN 90-277-1124-0
10.	Advances in European Geothermal Research, Strasbourg, March 1980	D. Reidel, Dordrecht (1980) EUR 6862 EN ISBN 90-277-1138-0
11.	Non Technical Obstacles to the use Solar Energy, Brussels, May 1980	Harwood Academic Publishers GmbH, Chur (1980), EUR 7003 EN
12.	Third E.C. Photovoltaic Solar Energy Conference, Cannes, October 1980	D. Reidel, Dordrecht (1981) EUR 7089 EN ISBN 90-277-1230-1
13.	Energy from Biomass Brighton, November 1980	Applied Science Publishers Ltd. London (1981) EUR 7091 EN
14.	Workshop on Medium Sized Photovoltaic Systems, Sophia Antipolis, 1980	D. Reidel, Dordrecht (1981) EUR 7090 EN

Other Publications

1. Energy Research and Development Programme Status Report, Martinus Nijhoff, The Hague (1977) ISBN 90-247-2059.

2. Energy Research and Development Programme, Second Status Report (1975—79) Martinus Nijhoff, The Hague (1979) ISBN 90-247-2220.9

3. Energy Models for the European Community, A. Strub (Ed.) IPC Science and Technology Press Ltd. Guilford (1979) ISBN 0-86103-011.7

4. Crucial Choices for the Energy Transition. CEC, EUR 6610EN (1980)

5. Atlas of Subsurface Temperatures in the European Community, Th. Schäfer GmbH, Tivolistrasse 4, D.3000 Hannover — 1 EUR 6578 EN

6. European Solar Radiation Atlas, W. Palz (Ed.) W; Grosschen, Verlay, Dortmund (1980) ISBN 3-8057-00637 EUR 6577 DA, DE, EN, FR, IT, NL

7. Energy for Biomass in Europe, W. Palz and P. Chartier (Eds.) Applied Science Publishers, Barking (1980) ISBN 0-85334-934-7

8. L'Energie Solaire au Service du Développement, CEC, Technique and Documentation, 11, rue Lavoisier, F — 75384, Paris Cedex 08 (1980)

9. Solar Houses in Europe, W. Palz and T. L. Steemers (Eds.) Pergamon, Oxford (1981) ISBN 0-08-026743.2

10. Passive Solar Architecture in Europe, Results of the First European Passive Solar competition, Ralph Lebens (Ed.) Architectural Press, London (July 1981).

If you have any concerns about our products,
you can contact us on
ProductSafety@springernature.com

In case Publisher is established outside the EU,
the EU authorized representative is:
**Springer Nature Customer Service Center GmbH
Europaplatz 3, 69115 Heidelberg, Germany**

Printed by Libri Plureos GmbH
in Hamburg, Germany